To all curious and imaginative children, may this book inspire you to explore the wonders of science and the world around you.

To Cuneyt and Feyza.

# Discovering The World Around Us

## Space Adventurer: A Journey Through Our Solar System

### A Kid's Guide to Solar System and Planets

Dr. Damon Polat is an enthusiastic physics and science teacher, school leader, teacher coach and human resource pro. He is passionate about inspiring children to be curious about the world around us. He believes that the world will be a better place when children feel the joy of learning science and use their knowledge for the benefit of all humanity.

To contact him, please email damonpolat@gmail.com or visit www.damonpolat.com

Copyright @ PLT Consulting and Management, LLC
All rights reserved. No part of this publication may be reproduced, stored or transmitted in any form, electronic, mechanical, or other without prior written permission of the owner.

Discovering The World Around Us

# Space Adventurer:
# A Journey Through Our Solar System

## A Kid's Guide to Solar System and Planets

by Dr. Damon Polat

## Table of Contents

| | | |
|---|---|---|
| Chapter 1 | Solar System | 1 |
| Chapter 2 | The Sun: Our Glowing Guardian | 5 |
| Chapter 3 | Mercury: The Nimble Messenger | 13 |
| Chapter 4 | Venus: The Mysterious Beauty | 19 |
| Chapter 5 | The Earth: Sweet Home | 25 |
| Chapter 6 | Mars: The Red Planet Next Door | 31 |
| Chapter 7 | Jupiter: The Mighty Gas Giant | 37 |
| Chapter 8 | Saturn: The Jewel of the Solar System | 43 |
| Chapter 9 | Uranus: The Tilted Ice Giant | 49 |
| Chapter 10 | Neptune: The Windy Blue Outpost | 55 |
| Chapter 11 | Dwarf Planets and Beyond | 61 |
| Chapter 12 | The Space Missions and Future Discoveries | 69 |
| Chapter 13 | Careers in Space Exploration | 77 |

The universe is huge. It includes billions of galaxies and each galaxy can contain billions of stars. This picture shows the Milky Way Galaxy and one of the bright spots is our star, the Sun.

## Chapter 1 Solar System

Welcome, young adventurers, to an exciting journey through space!

Together we'll explore the amazing family of celestial bodies we call the Solar System. On this fantastic journey, we'll visit the Sun, our planets, and other fascinating wonders to discover.

Our solar system is like a cosmic neighborhood in the vastness of the universe. At its heart lies the mighty Sun, a glowing ball of fire that brightens our day and warms our home, Earth. The Sun isn't only the brightest star in the sky, but also the most important member of our celestial family. Without the Sun, life on Earth wouldn't be possible!

Solar system is made up of the Sun, eight planets, many moons and some other objects move around it.

Did you know that there are eight planets orbiting the sun? Each of them is unique and full of secrets just waiting to be explored. We travel from the fastest planet, Mercury, to the farthest, Neptune. Along the way, we'll visit Earth's twin, Venus, and the Red Planet, Mars.

We'll also meet giant planets like Jupiter and Saturn, with their stunning moons and beautiful rings. And finally, we'll dive into the cold, distant worlds of Uranus and Neptune.

But that's not all! Our solar system is also home to countless other celestial objects such as asteroids, comets, and dwarf planets like the famous Pluto.

So, buckle up and get ready for an unforgettable adventure in the vastness of space! Let the journey begin!

Our star, the Sun is gigantic. It loses 4.7 million tons of its mass every second to emit heat and light so that we can have life on Earth.

## Chapter 2 The Sun: Our Glowing Guardian

Welcome to the center of our solar system, where our glowing guardian, the Sun, resides! This enormous ball of fire isn't only the brightest object in our sky, but also vital to the Earth.

Let's learn more about this glowing star that holds our cosmic family together.

Did you know that the Sun is so big that it could fit about 1.3 million Earths? That's right! It's a true giant making up 99.86% of the total mass of the solar system.

The Sun is really far away, about 93 million miles (150 million kilometers) from us. The heat and light of the Sun make life possible on our planet Earth.

The sun is very hot. Because of its high temperature, hydrogen atoms can transform into helium by nuclear fusion and generate enormous energy.

Sunlight can reach us in about 8.5 minutes!

The Sun is like a giant, fiery factory that generates energy through a process called nuclear fusion.

Deep in its core, the immense pressure and heat of the Sun cause hydrogen atoms to combine and form helium.

This process releases an incredible amount of energy in the form of light and heat that travels through space all the way to Earth.

You may be wondering what the Sun is made of. Well, it's mostly hydrogen (about 74%) and helium (about 24%), with tiny amounts of other elements like oxygen, carbon, and iron.

These beautiful light shows called aurora caused by Sun flares when it is very active.

The surface of the Sun, called the photosphere, has a temperature of about 9,932 degrees Fahrenheit. That's hotter than the hottest lava on Earth!

If you look up at the sky on a sunny day, you might see beautiful patterns of light called sunbeams.

These sunbeams are actually streams of charged particles called solar wind that emanate from the Sun and fill our solar system.

Sometimes the Sun becomes very active and releases huge bursts of energy called solar flares. These flares can cause beautiful light shows on Earth called auroras.

As we continue our journey through the solar system, remember that our mighty Sun is the source of life-giving energy for Earth and the force that holds our cosmic family together.

With its warm embrace and powerful light, the Sun is a beacon that lights our way as we explore the wonders of space.

So let's say goodbye to our glowing guardian and set sail for our first planetary destination: Mercury!

Mercury is closest to the Sun and slightly larger than our Moon.

## Chapter 3 Mercury: The Nimble Messenger

As we race through space, we reach our first planetary stop: Mercury, the planet closest to the Sun! Mercury is a small and fast world full of surprises. Let's explore this fascinating planet and uncover its secrets.

Mercury is the smallest planet in our solar system and only slightly larger than Earth's moon. If you could line up 18 Mercurys side by side, they'd be about as wide as Earth!

Mercury is a heavy, dense planet that, despite its small size, is composed mostly of rock and metal.

You may think that Mercury is the hottest planet because it's so close to the Sun.

Mercury has little atmosphere so it can't trap heat energy coming from the Sun. That's why the temperature on Mercury fluctuates a lot.

However, that title goes to Venus! But don't be fooled: Mercury still has blazing temperatures up to 800 degrees Fahrenheit during the day. That's hot enough to melt lead!

At night, however, Mercury's temperature drops dramatically, reaching as low as −290 degrees Fahrenheit.

Mercury has little atmosphere to trap heat, which is why its temperature fluctuates so much.

The lack of an atmosphere also means that there is no weather or wind on Mercury, and the planet's surface is littered with craters, just like our moon!

One of the most fascinating things about Mercury is how fast it orbits the Sun.

One year on Mercury takes 88 Earth days while one day on Mercury takes 59 Earth days. Which is the Planet Mercury on this picture?

Mercury takes only 88 Earth days to complete one orbit, making it the fastest planet in our solar system.

However, a day on Mercury is much longer than you might expect. It takes 59 Earth days for Mercury to rotate once on its axis. This means that a single day on Mercury is almost as long as two whole months on Earth!

As we say goodbye to the fast and mysterious planet Mercury, we prepare for our next adventure. We're about to visit Earth's sister planet, the dazzling and enigmatic Venus.

Get ready for a journey to a world of extreme heat and swirling clouds!

The Earth's sister planet, Venus is about the same size of our home, Earth, and second in distance to Sun after Mercury.

## Chapter 4 Venus: The Mysterious Beauty

We have now arrived at the second planet from the Sun, Venus, often referred to as "Earth's sister planet." Venus is a world of wonder and mystery. It's time to find out what makes this dazzling planet so beautiful and mysterious.

Venus is similar to Earth in size and composition, but that's where the similarities end. The atmosphere on Venus is thick and composed mostly of carbon dioxide, with clouds of sulfuric acid.

These clouds form a thick veil that hides the planet's surface from view. In fact, Venus is so cloudy that it reflects most of the sunlight that reaches it, making it the third brightest object in our sky after the Sun and the Moon!

Venus seems very bright when you look up at night because it has very thick clouds so it reflects the majority of the sunlight and it is closest to the Earth.

Although it's farther from the Sun than Mercury, Venus is the hottest planet in the solar system. The temperature on its surface is a blistering 870 degrees Fahrenheit, hot enough to melt lead! This extreme heat is due to the greenhouse effect, in which Venus' dense atmosphere absorbs heat from the Sun and raises the planet's temperature.

Venus has fascinating surface features, including mountains, valleys, and vast volcanic plains. The highest mountain on Venus, Maxwell Montes, is about 8.8 kilometers high, higher than Mount Everest, the highest mountain on Earth!

There are also more volcanoes on Venus than on any other planet in the solar system, over 1,600 of them.

A day on Venus is longer than its year! And it is the hottest planet, a whopping 870 degrees at its surface.

A day on Venus is unlike any other in our solar system. Venus rotates on its axis very slowly and in the opposite direction compared to most other planets, including Earth.

That is, the sun rises in the west and sets in the east on Venus! A day on Venus lasts 243 Earth days, while its year, the time it takes to orbit the Sun, is only 225 Earth days.

A day on Venus is therefore longer than her year!

As we leave behind the mysterious beauty of Venus, we return to our home planet, Earth.

Get ready to rediscover our unique and wonderful world and learn more about why Earth is the perfect place for life.

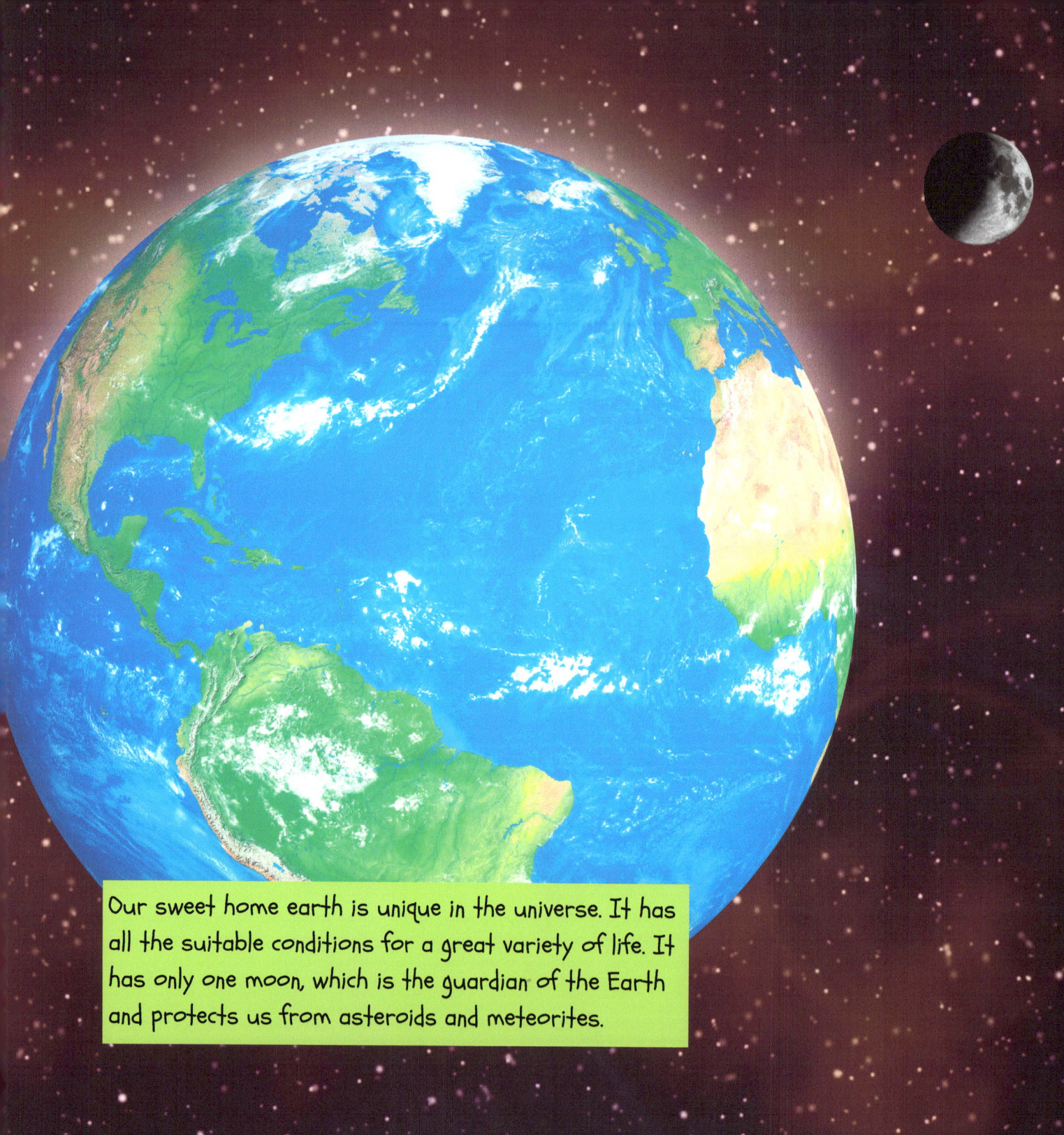

Our sweet home earth is unique in the universe. It has all the suitable conditions for a great variety of life. It has only one moon, which is the guardian of the Earth and protects us from asteroids and meteorites.

# Chapter 5 The Earth: Sweet Home

Welcome back to our planet, Earth, the third stone from the Sun and the only known place in the universe where life exists! Earth is a beautiful blue marble teeming with life and amazing natural wonders.

Let's rediscover our home and learn what makes it so unique.

Earth is often called the "Blue Planet" because about 71% of its surface is water. The vast oceans, seas, lakes and rivers are a vital resource on which life can thrive.

Water plays a crucial role in the Earth's climate, weather and water cycle, which includes evaporation, condensation and precipitation.

Earth is super beautiful because it provides many different habitats for animals and plants and is the only home in the big universe that we know, where everything grows and plays together.

Our planet has a unique atmosphere composed primarily of nitrogen and oxygen, with small amounts of other gasses such as carbon dioxide and water vapor.

The atmosphere protects life on Earth by keeping the temperature stable, blocking harmful solar rays, and burning most meteoroids before they reach the surface.

Earth hosts an incredible variety of ecosystems, from scorching deserts and lush rainforests to the icy polar regions and the mysterious depths of the oceans. These diverse habitats are home to millions of species of plants, animals, and microorganisms, making the Earth a living tapestry of life.

The surface of our planet isn't only made up of water and land, but is also divided into massive tectonic plates.

Although we can live safely on the surface of the Earth, its interior is very hot and full of lava.

These plates are constantly moving, albeit very slowly, and their movement causes the Earth's surface to change over time. This process is responsible for the formation of mountains, volcanoes and earthquakes.

The Earth takes 24 hours to rotate once on its axis, giving us the day and night cycle. It also takes 365.25 days to go around the sun, which makes up our year.

The tilt of the Earth's axis in its orbit around the Sun is responsible for the seasons, as different parts of the planet receive different amounts of sunlight throughout the year.

Now let's continue our journey through the solar system and head to our closest neighbor, the fascinating Red Planet Mars!

Scientists have discovered polar ice caps that change with the seasons. They claim that there were water channels on Mars.

## Chapter 6 Mars: The Red Planet Next Door

We have reached the fourth planet from the Sun, Mars, our fascinating next-door neighbor! Because of its rusty red appearance, Mars is often called the "Red Planet."

Let's explore this captivating world and learn why it could be a potential future home for humans.

Mars is about half the size of Earth, but it has some truly remarkable features. Its surface is rocky and dusty, with vast plains, enormous volcanoes, and deep canyons.

Olympus Mons, the tallest volcano on Mars, is an astonishing 13.6 miles high, making it the tallest known volcano in the entire solar system!

Our neighbor, the "Red Planet" Mars, seems to be the closest place that is habitable for humans.

Mars also has the longest canyon, Valles Marineris, which is over 2,485 miles long, about as long as the United States is wide!

The atmosphere on Mars is thin and consists mainly of carbon dioxide, with traces of nitrogen and argon. Because of the thin atmosphere, temperatures on Mars can be quite extreme. They range from a cool -195 degrees Fahrenheit on winter nights to a relatively mild 68 degrees Fahrenheit on summer days.

One of the most intriguing things about Mars is the evidence of water on its surface. Although liquid water cannot exist on Mars today because of low atmospheric pressure, scientists have discovered evidence of ancient riverbeds, polar ice caps and underground water reserves.

Mars has two small, irregularly shaped moons named Phobos and Deimos.

This raises the possibility that suitable conditions for life existed on Mars.

Mars has two small, irregularly shaped moons named Phobos and Deimos. These moons are believed to be captured asteroids.

Scientists and engineers are developing new technologies and missions to explore the Red Planet as we learn more about Mars and its potential for human life. One day, you could be among the future generation of astronauts who call Mars their second home!

As we say goodbye to Mars, it's time to move on to the next stage of our solar system adventure. Get ready to enter the realm of gas giants, starting with the largest and most magnificent of them all, Jupiter!

Jupiter is by far the largest planet in our solar system. Its mass is almost twice as large as that of all other planets together.

## Chapter 7: Jupiter: The Mighty Gas Giant

We have now arrived in the realm of the gas giants, and our first stop is the largest and most magnificent planet in the solar system: Jupiter!

Jupiter is a colossal world with a powerful presence. Let's dive into the fascinating features of this mighty gas giant.

Jupiter is so enormous that over 1,300 Earths would fit inside it! Despite its massive size, Jupiter is composed mostly of hydrogen and helium, like the Sun.

It has no solid surface, but layers of thick clouds and swirling storms. One of Jupiter's most striking features is the Great Red Spot, a gigantic storm that has been raging for at least 400 years.

Jupiter has at least 79 known moons, the four largest of which are called the Galilean moons: Io, Europa, Ganymede, and Callisto.

This colossal storm is so large that it could swallow three Earths side by side! Jupiter's atmosphere is also known for its stunning layers of color, created by the planet's fast winds and various cloud layers.

Jupiter's strong gravity affects not only its atmosphere, but also its numerous moons. Jupiter has at least 79 known moons, the four largest of which are called the Galilean moons: Io, Europa, Ganymede, and Callisto.

These moons were discovered by the famous astronomer Galileo Galilei in 1610 and are fascinating worlds in their own right.

Ganymede, the largest moon in the solar system, is even larger than the planet Mercury!

Europa, the icy moon of Jupiter, could be one of the most suitable places for life outside the earth.

On the other hand, a vast ocean of liquid water is believed to lie beneath Europa's icy surface, making it a potential candidate for extraterrestrial life.

Jupiter takes about 10 hours to make one revolution, making it the fastest rotating planet in the solar system. This rapid rotation causes the planet to bulge at its equator and flatten at its poles.

Jupiter takes about 12 Earth years to orbit the Sun. One year on Jupiter therefore corresponds to 12 years on Earth!

As we bid farewell to mighty Jupiter, we continue our journey to the next gas giant in our celestial family: the spectacular ringed planet Saturn. Be amazed by its stunning beauty and fascinating secrets!

Saturn is known for its stunning rings and captivating appearance. It's first discovered by Galilei in 1610.

# Chapter 8 Saturn: The Jewel of the Solar System

Our next destination is the stunningly beautiful ringed planet Saturn. Saturn is often referred to as the "jewel of the solar system." It's the sixth planet from the Sun and the second largest in our celestial family.

Let's explore this fascinating world and learn about its impressive features.

Saturn is composed mostly of hydrogen and helium, much like its neighbor Jupiter. It's not quite as big as Jupiter, but it's still a giant that can fit about 764 Earths!

Saturn has an interesting property: it has a lower density than water. This means that if a bathtub were big enough, Saturn would float!

Saturn is known for its stunning rings and captivating appearance. We cannot walk on Saturn's rings because they aren't solid structures. The rings are made up of countless individual particles, mostly ice, ranging from tiny grains to large chunks several meters in diameter.

Saturn's most famous feature is its dazzling ring system. These rings are made up of billions of particles, ranging from tiny grains of dust to huge chunks of ice and rock, orbiting the planet.

The rings are incredibly thin, only about 33 feet thick. Yet they span 175,000 miles, or about three-quarters of the distance between Earth and the Moon!

Saturn has an impressive collection of moons, with at least 83 known satellites. The largest moon, Titan, is even larger than Mercury and has a thick atmosphere composed mostly of nitrogen.

It's the only moon in the solar system with a thick atmosphere, and it's believed that there are lakes and rivers of liquid hydrocarbon on its surface.

Saturn has 83 moons. The largest is called Titan, is larger than Mercury, and has a thick atmosphere and liquid lakes on its surface.

A day on Saturn is relatively short, as the planet takes only 10.7 hours to make one rotation. However, its journey around the Sun is much longer: a Saturnian year lasts about 29.5 Earth years.

This means that a person who is 30 years old on Earth would be only a little over 1 year old on Saturn!

Now that we have left behind the enchanting realm of Saturn and its breathtaking rings, let's head into the depths of the solar system.

Our next stop is the ice giant Uranus, a mysterious blue world with a fascinating history. Get ready to uncover the secrets of this distant planet!

Uranus has a bluish color because its atmosphere contains methane gas.

## Chapter 9 Uranus: The Tilted Ice Giant

On our journey through the solar system we now come to the seventh planet from the sun, the mysterious ice giant Uranus.

This fascinating blue world has a unique history and some amazing features. Let's uncover the secrets of this distant and cool planet.

Uranus is a giant, gaseous planet composed mainly of hydrogen, helium, and a mixture of ices, including water, ammonia, and methane. The methane in its atmosphere gives Uranus its distinctive blue color.

It's not as big as Jupiter or Saturn, but Uranus is still large enough to hold about 63 Earths!

One of the most unusual things about Uranus is its extreme tilt. Unlike other planets, Uranus turns on its side and tilts its axis 98 degrees.

This means that the planet appears to roll in its orbit instead of spinning like a top. Scientists believe that a massive collision with another celestial body may have caused this strange tilt.

Uranus has 27 known moons, most of which are named after characters in the works of William Shakespeare and Alexander Pope.

The five largest moons – Miranda, Ariel, Umbriel, Titania, and Oberon – are icy worlds with unique geological features such as huge canyons, towering cliffs, and vast, apartment plains.

Uranus has a unique axial tilt that causes it to turn on its side. This causes extreme seasonal variations on Uranus. Each pole experiences 42 years of uninterrupted sunlight in summer, followed by 42 years of total darkness in winter.

A day on Uranus is quite short: the planet takes only about 17.2 hours to make one rotation. Its journey around the sun, however, is much longer: a Uranus year lasts a full 84 Earth years!

This means that the poles of the two planets experience 42 years of uninterrupted sunlight, followed by 42 years of darkness.

As we bid farewell to the tilted ice giant Uranus, we make our way to the outermost planet in our solar system, the mysterious and enchanting Neptune.

Get ready for a journey to a world with strong winds, freezing temperatures, and a captivating blue hue!

Neptune is the eighth planet from the Sun and is considered an ice giant, similar to its neighbor Uranus.

# Chapter 10 Neptune: The Windy Blue Outpost

Our adventure through the solar system brings us to the outermost planet, enchanting and distant Neptune. This icy blue world is a realm of strong winds, extreme temperatures, and mesmerizing beauty.

Let's explore the captivating features of this distant outpost.

Neptune is the eighth planet from the Sun and is considered an ice giant, similar to its neighbor Uranus. It's composed primarily of hydrogen, helium, and a mixture of ices, including water, ammonia, and methane.

The methane in Neptune's atmosphere absorbs red light and reflects blue light, giving the planet its striking blue color.

Triton is the largest moon of Neptune. It's unique among the large moons of our solar system because it orbits Neptune in the opposite direction to the planet's rotation.

Neptune is about four times the size of Earth, but its distance from the Sun makes it much colder. The average temperature on Neptune is a frigid -353 degrees Fahrenheit, making it one of the coldest places in the solar system.

Despite its cold and calm appearance, Neptune has the strongest winds in the solar system, reaching up to 1,304 miles per hour! These strong winds whip up huge storms, including a massive dark storm, the Great Dark Spot, which is similar in size to Earth.

Neptune has 14 known moons, of which Triton is the largest and most fascinating. Triton is unique in that it orbits Neptune in the opposite direction of the planet's rotation. It was likely captured by Neptune's gravity and was once a wandering object in the solar system.

A day on Neptune is relatively short, as the planet takes about 16 hours to make one rotation. Its journey around the sun, however, is incredibly long: A Neptunian year lasts 165 Earth years! Since its discovery in 1846, Neptune has completed only one orbit around the Sun.

To conclude our exciting journey through the solar system, we explored the captivating worlds that make up our celestial family. Each world offers unique wonders and mysteries, from the fiery sun and rocky planets to the majestic gas and ice giants.

Our adventure may be over, but the universe is vast and there is always more to explore and discover. Keep your eyes on the stars and your imagination alive, because the adventure never really ends!

Once considered the ninth planet, Pluto was reclassified as a dwarf planet in 2006.

# Chapter 11 Dwarf Planets and Beyond

As we continue our adventure through the solar system, we move to the edge of the system where we discover a fascinating group of celestial bodies known as dwarf planets.

These small, unique worlds offer exciting insights into the formation of our solar system and the mysteries that lie beyond.

Dwarf planets are celestial bodies that orbit the Sun and are large enough to have a round shape due to their gravity. Yet they haven't cleared their orbits of other debris. Let's explore some of the best known dwarf planets and their fascinating properties.

Dwarf planets Pluto, Eris, Haumea, Makemake, Seres and their moons.

**Pluto:** Once considered the ninth planet, Pluto was reclassified as a dwarf planet in 2006. This small, icy world has a heart-shaped region called the Tombaugh Regio, named after its discoverer, Clyde Tombaugh. Pluto has five known moons, the largest being Charon, which is nearly half the size of Pluto itself!

**Eris:** Eris is a dwarf planet that is even more distant than Pluto. The discovery of Eris, which is larger than Pluto, led to a redefinition of what a planet is. Eris has a single moon named Dysnomia.

**Haumea:** This elongated dwarf planet is unique because of its unusual shape caused by its rapid rotation. Haumea is known to have two moons, Hi'iaka and Namaka, and is also surrounded by a thin layer of ice.

Ceres is the largest object in the asteroid belt between Mars and Jupiter and is the only dwarf planet in the inner solar system.

Makemake: Makemake is another icy dwarf planet with no known moons. It has a reddish-brown color and is located in the region of the solar system known as the Kuiper Belt.

Ceres: Ceres is the largest object in the asteroid belt between Mars and Jupiter and is the only dwarf planet in the inner solar system. It's known for its bright spots, which are believed to be deposits of salt or ice.

Beyond the dwarf planets lies the vast Oort Cloud, a distant region of icy objects thought to be the source of long-period comets. These comets have very long orbits and can take thousands of years to make a single trip around the Sun.

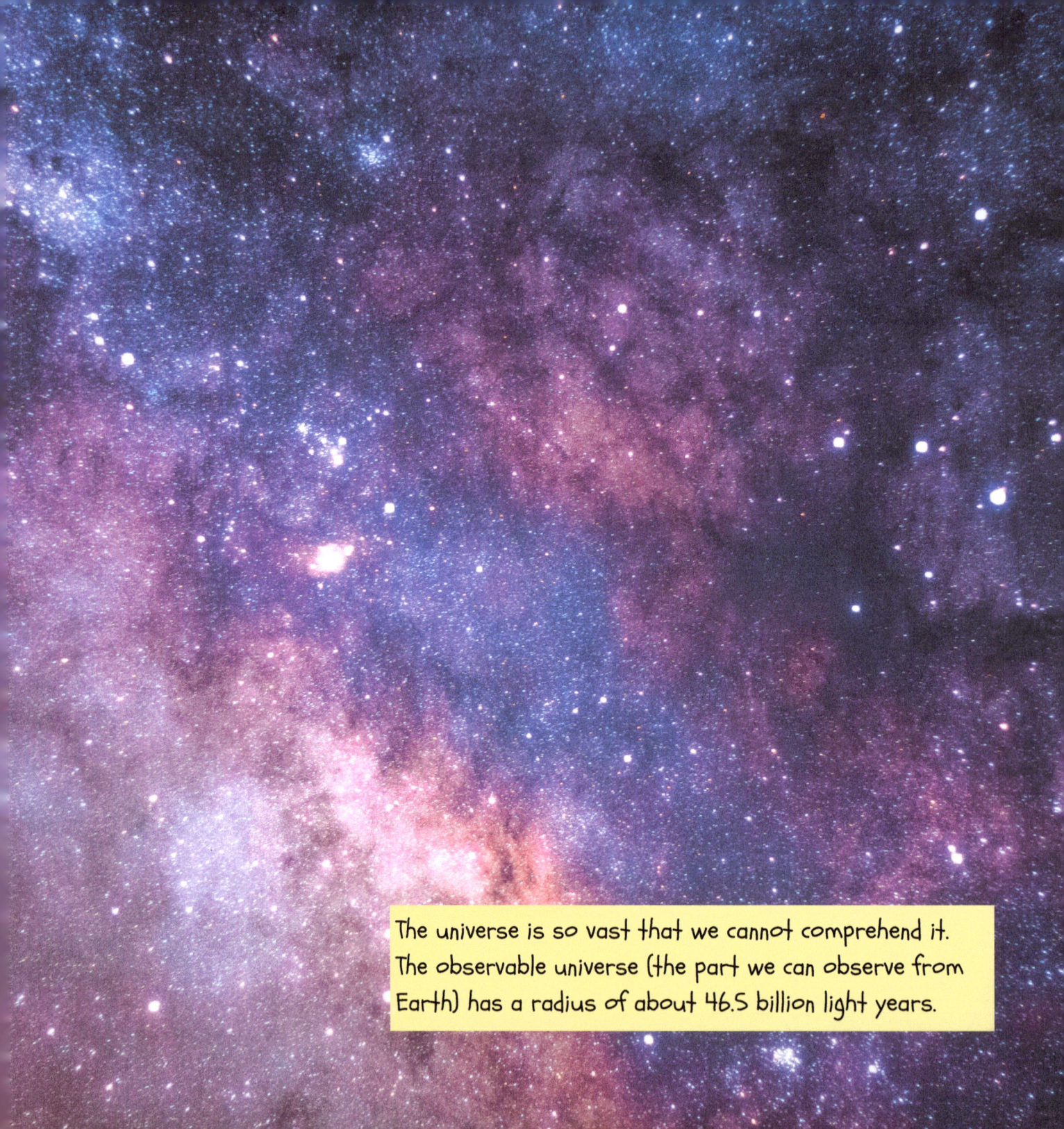

The universe is so vast that we cannot comprehend it. The observable universe (the part we can observe from Earth) has a radius of about 46.5 billion light years.

Our exploration of the dwarf planets and the outer reaches of the solar system illustrates the vastness and diversity of worlds that make up our celestial neighborhood.

At the end of our adventure, remember that the universe is full of wonders waiting to be discovered. The thrill of exploration is always just a thought away. Keep looking up at the night sky and let your curiosity guide you on your cosmic journey!

As you delve into the mysteries of the cosmos, you'll find that science isn't only fascinating, but also an inexhaustible source of inspiration and excitements. So, young explorers, don't be afraid to ask questions, dream big, and imagine the endless possibilities that await you in the universe.

The International Space Station (ISS) is a large, habitable spacecraft and research facility that orbits the Earth at an altitude of about 400 kilometers.

## Chapter 12 Space Missions and Future Discoveries

As we reflect on our adventure in the solar system, it's important to acknowledge the incredible space missions and dedicated scientists who have helped us unlock the mysteries of our celestial neighborhood.

In this final chapter, we learn more about some groundbreaking missions and the exciting future of space exploration.

**Voyager 1 and 2:** Launched in 1977, these two space probes have provided us with stunning images and valuable data about gas giants and their moons. Today, they continue their journey into interstellar space, making them the furthest man-made objects from Earth.

Several rover robots have been sent out to explore the surface of Mars.

**Mars rovers:** Several rover robots, including Spirit, Opportunity, Curiosity, and Perseverance, have been sent out to explore the surface of Mars. These missions have provided invaluable information about the geology of Mars, its atmosphere, and the potential for life.

**Cassini-Huygens:** This joint mission from NASA and the European Space Agency (ESA) provided unprecedented insights into Saturn, its rings and its moons, particularly Titan. The mission ended in 2017 when the Cassini spacecraft intentionally crashed into Saturn's atmosphere.

**New Horizons:** Launched in 2006, the spacecraft made a historic flyby of Pluto and its moons in 2015, giving us our first up-close look at these distant worlds. New Horizons then continued to explore objects in the Kuiper Belt.

The James Webb Space Telescope is a large, infrared-optimized space telescope with advanced equipment for studying galaxy formation. Its large mirror is 6.5 meters in diameter.

**James Webb Space Telescope:** This powerful telescope was launched in December 2021 to observe the universe in the infrared. It helps us study the formation of galaxies, stars and planets and search for signs of life on exoplanets.

Some of the upcoming missions and concepts include:

**The Artemis Program:** This program aims to return humans to the Moon by 2024 to create a sustainable lunar presence and use it as a springboard for future missions to Mars.

**Mars Sample Return:** A mission to collect samples from Mars and return them to Earth for analysis and to provide valuable information about the geology of the Red Planet and the potential for life.

**Europa Clipper:** A planned mission to explore Jupiter's moon Europa, which is believed to have an ocean beneath its icy crust. The goal of this mission is to explore the moon's potential habitability and search for signs of life.

As we end our journey through the solar system, remember that the spirit of exploration and curiosity drives us to discover the mysteries of the cosmos.

As you look up at the stars, dream big and let your imagination run wild, for you're the future generation of explorers, scientists, and dreamers who will continue to unlock the secrets of our universe.

The adventure has just begun!

An astronaut in outer space left the International Space Station to perform a spacewalk to test new technologies.

## Chapter 13: Stargazers and Astronauts: Careers in Space Exploration

As our adventure through the solar system draws to a close, you may be wondering how you can become a part of the exciting world of space exploration.

In this final chapter, we'll introduce you to some careers that can help you turn your passion for the cosmos into a lifelong journey of learning and discovery.

**Astronomer:** Astronomers explore celestial objects such as stars, planets, galaxies, and more. They use telescopes, advanced software, and mathematical models to analyze data and make new discoveries about the universe.

The first space shuttle, Columbia, launched on April 12, 1981, and astronauts John Young and Robert Crippen spent 54 hours in Earth orbit and returned home safely..

Astrobiologist: Astrobiologists look for signs of life beyond Earth by studying extreme environments on our planet and exploring the possible habitability of other worlds such as Mars or Europa.

Aerospace Engineer: These engineers design, build, and test spacecraft, rockets, and satellites. They play a critical role in the development of new space and exploration technologies.

Astronaut: Astronauts train for years to participate in manned space missions. They conduct scientific research, maintain spacecraft and explore celestial bodies such as the moon or Mars.

The International Space Station (ISS) was first launched into orbit in 1998 and can often be seen from Earth with the naked eye. It orbits at an altitude of 250 miles above Earth and travels at a speed of 17,500 miles per hour..

**Planetary geologist:** Planetary geologists study the geology of planets, moons, and other celestial bodies. They analyze data from space probes and telescopes to understand the history and evolution of these worlds.

**Astrobiologist:** Astrobiologists look for signs of life beyond Earth by studying extreme environments on our planet and exploring the possible habitability of other worlds such as Mars or Europa.

**Science Communicator:** Science communicators convey the fascination of space exploration to the public. They may write articles, create educational content, give lectures, or work in planetariums and science museums.

**Space weather scientists:** These scientists study solar activity, such as solar flares and coronal mass ejections, and their effects on Earth's magnetosphere and satellites. They help protect our technology from space weather events.

To pursue a career in space exploration, a solid foundation in science, technology, engineering, and mathematics (STEM) is essential. A degree in astronomy, physics, chemistry, biology, and computer science can give you the skills you need to embark on a journey through the cosmos.

As you explore and learn about the wonders of the solar system and the universe, remember that curiosity, determination, and a spirit of discovery are key to unlocking the mysteries of the cosmos. The sky isn't the limit, it's only the beginning of your journey!

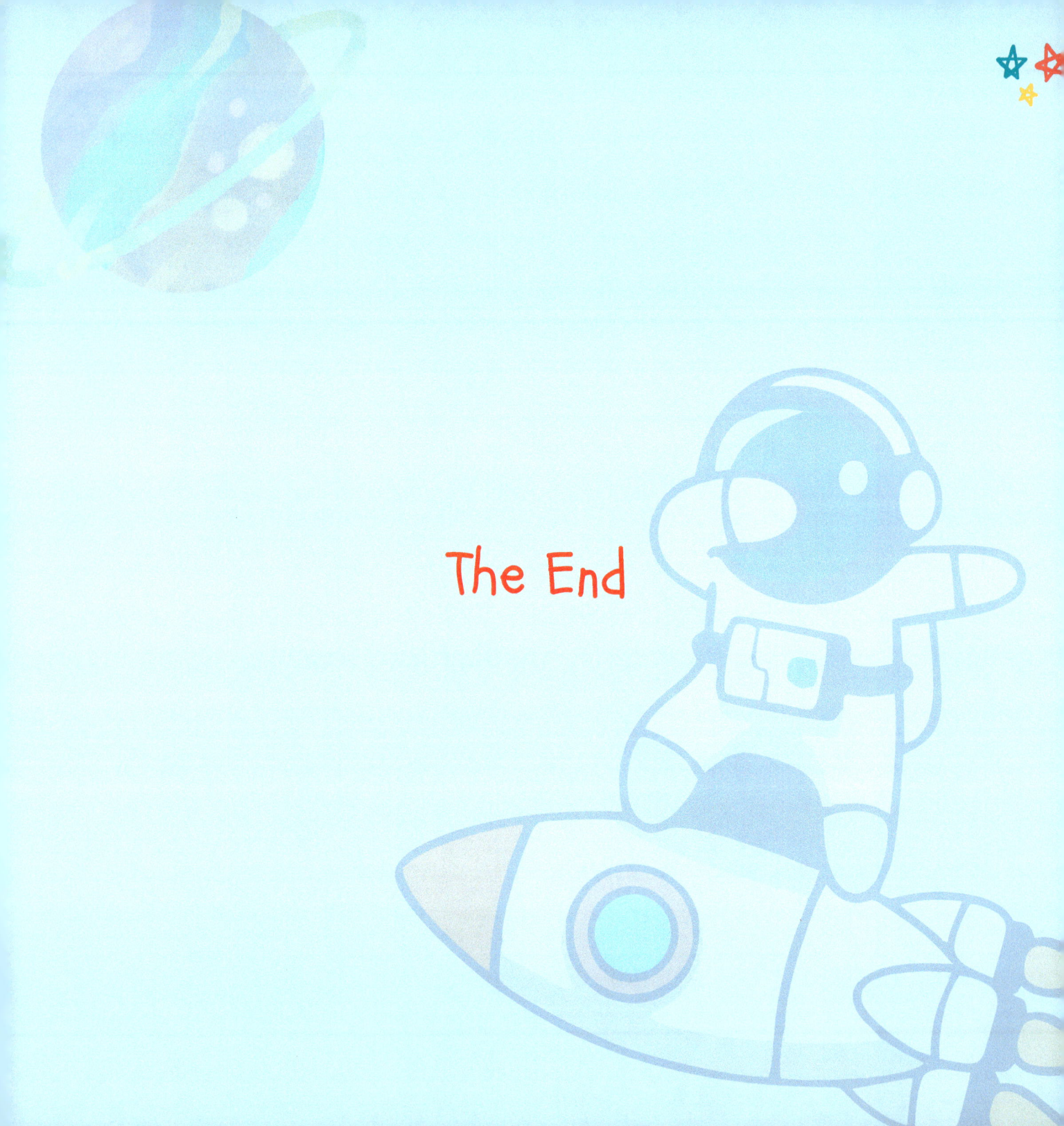

www.ingramcontent.com/pod-product-compliance
Lightning Source LLC
Chambersburg PA
CBHW051156220526
45473CB00003B/792